KNOWING MICROBES

A DIVE INTO THE FASCINATING WORLD OF MICROORGANISMS

KNOWING MICROBES

Copyright © Amina Ahmed El-Imam (Ph.D.) 2021

Illustrations & Design: Mackh Visuals & Amina Ahmed El-Imam

Published by: Amina Ahmed El-Imam

All Rights Reserved. No part of this publication may be reproduced, distributed or transmitted in any form or by any means including photocopying, recording or other electronic digital or mechanical methods, without the prior written permission of the author, except in the case of brief quotations embodied in critical reviews and certain other non-commercial uses permitted by copyright law. For permission requests, write to the author at knowingmicrobes@carnationicl.com.

This publication is designed to provide accurate and authoritative information in regard to the subject matter covered. It is sold with the understanding that the author is not engaged in rendering medical or other professional service. If medical advice or other expert assistance is required, the services of a competent professional should be sought.

Any brand names and product names used in this book are trademarks, registered trademarks or trade names of their respective holders. The author is not associated with any product or vendor in this book.

Knowing Microbes/Ahmed El-Imam, Amina.

 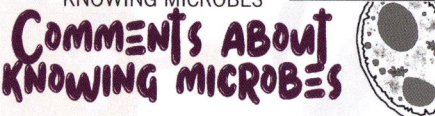

"Non-microbiologists can understand it"

I had the honour and privilege to review *Knowing Microbes* authored by Dr. Amina Ahmed El-Imam. The author's direct approach, made it such that non-microbiologists can comprehend the book's contents. The section on microbial classification included molecular taxonomy which gives the book exceptional reading value as most authors limit themselves to traditional taxonomy. This book also highlights the negative and positive relationships existing between microbes, man and the environment, and microbes' roles in bio-economic development.

I congratulate Dr. El-Imam for finding time to write the book, and I recommend it to all.

- Prof. M.B. Yerima, *FNSM, FBSN*

President, Nigerian Society for Microbiology (2018 – date)

President, Biotechnology Society of Nigeria (2021 – date)

""The author has laid bare our unseen allies"

Microbiology is the youngest arm of the life sciences, so it is relatively unknown even despite COVID-19. Because microbes are invisible, teaching about them to beginners can be difficult. This book, *Knowing Microbes*, presents a profound innovation and a great effort at introducing the subject at the foundation level.

The author has laid bare our unseen allies for all to see and study. It is lucidly written, fascinating, easy to read and comprehend as it is beautifully illustrated. I recommend it to those willing to extend their frontier of knowledge.

- Prof Albert Olayemi

Professor of Microbiology,

Former Deputy Vice-Chancellor,

University of Ilorin, Ilorin Nigeria.

"A great way to introduce kids to microbes...and science!"

Knowing Microbes by Dr. Amina Ahmed El-Imam is a great book for kids to appreciate microbes – what they are, what they do, and how they are relevant to us as humans. In her characteristic manner, Dr. El-Imam has made it fun-reading while learning quite a lot.

Furthermore, what Dr. El-Imam has done goes beyond her field of microbiology but extends to national development as no nation thrives without building on science, technology, and innovation (STI). Nigeria has been witnessing a decline in STI and one major reason is that fewer and fewer scientists are being raised. It has been found in research that if children will be interested in science, it is best to happen even while they are in elementary school. They can then grow up to pursue a distinguished career in the sciences.

- Dr. M. Oladoyin Odubanjo

Executive Secretary, The Nigerian Academy of Science.

Chair, International Network for Government Science Advice,

Africa Chapter (INGSA-Africa).

"This book is handy while the world is battling many infectious diseases"

Knowing Microbes gives a basic description of fundamentals to the world of microbes in six main sections. The book is designed and written in simple language for easy comprehension by readers. A book of this nature comes in handy at a time like this when the world is battling many infectious diseases. The ongoing COVID-19 pandemic and Ebola outbreak cannot be forgotten quickly. Although microbes are infamous, the book provided information on some beneficial uses of microorganisms especially in food preparations and health.

Knowing Microbes is ideally suited for secondary school pupils and non-science majors. Appropriate illustrations and examples are also provided to further make for interesting reading and application of knowledge to happenings around us. I, therefore, recommend this book for every family.

Adenike Temidayo Oladiji, FAS

Professor of Biochemistry

Dean, Faculty of Life Sciences,

University of Ilorin, Ilorin, Nigeria.

"Stimulating, captivating and definitely educative"

It is my pleasure to associate with this wonderful effort by Dr. Amina Ahmed El-Imam to make "the invisibles" known to young and curious minds. I have had the privilege of going through the scholarly work and I found it interesting, stimulating, captivating and definitely educative. I am particularly enthused because although it was originally targeted at children, it is presented in such a manner that it can very well serve as an introductory material for learning about microbes at higher levels. It takes a peep into the historical development of microbiology from the pre-Antonie van Leeuwenhoek era till present time. The author succinctly presents microbes as real things we live with daily and not just some abstract things talked about in science class. She aptly presented the different types of microbes and how they influence human activities and the biosphere.

I heartily congratulate Dr. Amina Ahmed El-Imam for her contribution in enhancing the learning of microbiology; it is a remarkable contribution towards *Knowing Microbes.*

- Prof. K.I.T. Eniola, *FNSM*

Country Ambassador (Nigeria), American Society for Microbiology

President, Nigeria Society for Microbiology (NSM) 2014 - 2018.

"Communicating science to non scientists is a specialist skill and Dr Ahmed El-Imam has done an excellent job"

The role of the infinitely small in nature is infinitely large' - Louis Pasteur. This quote captures the discipline of microbiology, the study of microbial life and the relationships between microbes and other organisms. This awesome project by Dr El-Imam introduces microbiology to young learners in a way that captures attention and inspires curiosity. Communicating science to non scientists is a specialist skill and Dr Ahmed El-Imam has done an excellent job. Ensuring that the discipline of microbiology can benefit from human diversity, it is important we provide opportunities for the next generation of scientists to become fascinated with our subject and that we catch them young!

Dr. Amara Anyogu,

School of Biomedical Sciences,

University of West London, United Kingdom.

"Knowing microbes deserves a visible space in your library"
Knowing Microbes is about microorganisms, the study of which is known
as microbiology. It explicitly explains what microbes are and how they impact everyone. Knowing Microbes provides a detailed introduction to the microbial world and how we can use this knowledge to improve society.
Microbes are ubiquitous, found even in extreme environments, and Dr. El-Imam provides a simplified discussion about these extremophiles. This book discusses the ability of microbes to cause disease in humans and the different applications of microbes in the food and pharmaceutical industries. Knowing Microbes extends the knowledge of the reader to understand and appreciate the contributions of microbes to society and the world at large. It ends with a glossary of terms used in it.
I hope you find this book on knowing microbes as exciting and as useful as I did, as it deserves a visible space in your library.

- Professor Benjamin Ewa Ubi
President, Biotechnology Society of Nigeria (BSN)
2016 – 2021.

"Fascinating yet detailed book"

Knowing Microbes gives a general overview of microbiology. It is simple and fascinating enough for the targeted reader, yet detailed enough to provide an overview. It provides a very fascinating way of linking the ancient with modern. *Knowing Microbes* will not only serve its young audience, but it will also serve as a reminder for experienced microbiologists. Well done to the author.

- Dr. Bolanle K. Saliu

Head, Microbiology Department, Faculty of Life Sciences,

University of Ilorin, Ilorin, Nigeria.

"Uses simple language to explain complex concepts"

Knowing Microbes uses simple language to explain complex concepts by providing a comprehensive view of the fascinating world of microbes. It elucidates the harmful effects of microbes and explains the oft-ignored beneficial effects of microbes. *Knowing Microbes* will undoubtedly stimulate curiosity in young readers about microbes. Our pupils at AIS will certainly be learning from this well-researched book.

"Knowing Microbes: an invaluable introduction into microbiology."

- Mr. Abdulgafar Abdulraheem

Head of Schools,

Aderoju International Schools (AIS),
Ilorin, Nigeria.

"I was captivated..."

After reading few pages of *Knowing Microbes*, I was captivated to complete it. The book gives a much deeper understanding of microbiology as it inculcates indispensable knowledge that brings about a shift in understanding the basis of microbiology and microorganisms. It is written in a precise and concise language which makes it easy to understand. This book suffices as a quick and reliable reference, as well as worthy legacy for generations of students and practitioners of microbiology and other related fields of study. It will also serve as a guide to children, adults, and students of medical microbiology.

Dr. Amina Ahmed El-Imam has demonstrated her in-depth knowledge of microbiology and microorganisms. I must also say that she is a great writer who has simplified the concept of microbiology even to the understanding of a layman.

Knowing Microbes is a summary of many microbiology books, and its beauty is that it tells a story of the existence, activities and impact of these unseen creatures. This book is highly recommended to everyone who desires to have a slice of knowledge about microorganisms.

The author needs to be congratulated on this remarkable piece of work.

- **Maryam Aminu (Ph.D.)**

Professor of Microbiology (Virology),

Head, Department of Microbiology and Deputy Dean,

Faculty of Life Sciences, Ahmadu Bello University, Zaria.

Senior Fulbright Scholar-2015 and UNESCO L'Oréal Fellow-2004.

KNOWING MICROBES

TABLE OF CONTENTS

Comments about Knowing Microbes	i
Contents	vii
Foreword	viii
Introduction	1
History of microbiology	6
Observing microorganisms: Microscopy	19
Classification of microorganisms	24
How microorganisms move	34
Microorganisms are a part of us!	37
Microorganisms in extreme environments	50
Conclusion	53
Glossary	55

FOREWORD

It gives me great pleasure to write the foreword of *Knowing Microbes* by Dr. Amina Ahmed El-Imam. I have been involved in Microbiology research and teaching for 27 years. I am the President of both the Nigerian Society for Microbiology (NSM) and Biotechnology Society of Nigeria (BSN). The author is an active NSM member and an intellectual of repute. *Knowing Microbes* is a product of her passion for microbiology, hard work and intellectual disposition.

The book is indeed broad-based yet straightforward. Chapter one gives a comprehensive and up to date history of microbiology; I have never come across a book that gives a better account than *Knowing Microbes*. Chapter two discusses the tool of our trade, the microscope, and microscopy. The third chapter focuses on microbial classification, while the fourth chapter discusses motility. In Chapter five, the importance of our microbiome was discussed and lastly, Chapter six discusses microbes that grow in extreme environments.

Nowadays, when (re-)emerging infections are ravaging humanity, there is a need to better understand microbes. The global economic devastation caused by COVID-19 is still evident and this book, highlights how microbes can be tools for economic recovery. This book is suitable for students, teachers, policymakers, health professionals, industry experts and other interested groups.

Congratulations to Dr. El-Imam for this scholarly feat. I will like to emphatically recommend this book to all microbiologists and non-microbiologists alike.

- **Prof. M.B. Yerima** FNSM, FBSN
President,
Nigerian Society for Microbiology and
Biotechnology Society of Nigeria (January, 2022).

INTRODUCTION

INTRODUCTION

Microorganisms, also called microbes, are living things that exist all around us but which we are unable to see! You may be wondering: "how is this possible?" Well, it's because they are really, really tiny, so we need to utilize some special equipment to view them. The term microorganism is derived from Greek: *mikros* meaning "small", and *organismos* meaning "organism", while microbe is from *mikros* and *bios* meaning "life". You may have previously also referred to microorganisms as bugs, germs, and so on.

For most of the existence of life on Earth, microbes were the only life forms. Researchers identified bacteria in amber and salt samples up to 250 million years old, before the dinosaurs' time! This finding showed that the physical structure of microorganisms has not changed much since the Triassic period. Like humans and animals, microbes also have genetic material in the form of deoxyribonucleic acid or DNA which they can transfer by various means (Figure 1).

There are various groups of microorganisms, including archaea, bacteria, and fungi, which all have unique characteristics. The study of these microorganisms is called microbiology. Microorganisms are ubiquitous, which means they are found everywhere. We encounter them in soil, air, water, and even on our bodies!

LOVE IN SHARING?

Did you know that bacteria can transfer their genetic material among each other? There are three main ways through which this can happen:

In **conjugation,** bacteria can transfer genetic material by contact or sometimes through a tube-like connection known as a pilus between two cells.

In **transformation,** a bacterium can pick up naked DNA floating in its environment and stitch it up into its own DNA

Lastly, in **transduction,** viruses can pick up DNA from one bacterium and deposit it in another!

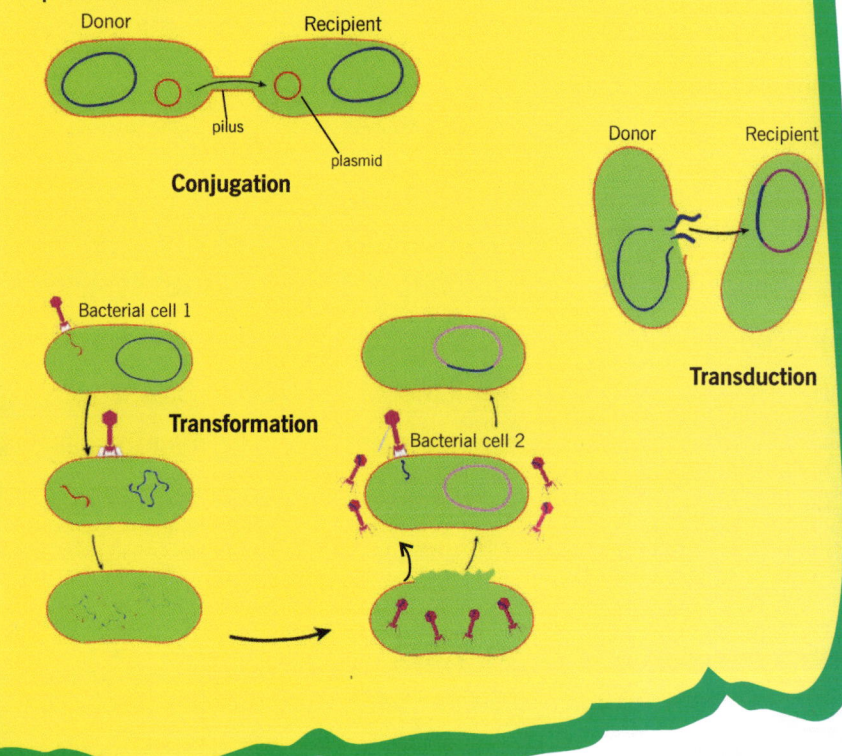

Figure 1. Genetic material transfer in bacteria

Even though we cannot see them with our eyes, we can see and even feel their activities around us. Microbes are involved in almost all aspects of human life, such as food production, spoilage, and decay, and making us ill. Thus, it is essential to understand them, protect ourselves from them, and even learn to exploit them for humans' benefit.

History of Microbiology

HISTORY OF MICROBIOLOGY

Microbiology as an area of biology helps us understand these tiny organisms, how they function, and how they interact with us and other living things. Did you know that microbiology has been existing in one way or another for millennia?

Saddle up; we are going for a long ride.

350 B.C.: Spontaneous generation was a common theory where people believed that life forms just erupted from non-living substances like air, decaying flesh, etc. Aristotle clarified this theory. It suggested that life came from no life; living things formed on their own from non-living things.

11th Century: Abu Ali Ibn Sina (Avicenna) proposed a basic form of the germ theory of disease.

13th Century: People knew that "invisible things" caused disease and spoilage, but no one knew what those things were.

14th Century: Lisan Al-Din Ibn Al-Khatib explored the idea that the transmission of the deadly plague was through contagion.

KNOWING MICROBES

15th Century: Akshamsaddin Muhammad (Figure 2), an Ottoman scientist and religious scholar, first mentioned the microbe in his book titled Maddat ul-Hayat (The Material of Life). He also developed a theory about germs, where he likened them to seeds.

Figure 2: Akshamsaddin Muhammad

16th Century: Girolamo Fracastoro, an Italian scholar, suggested that contagion could pass from one thing to another. He proposed that tiny particles or "spores" caused epidemics.

1660s: Dutch draper Antonie von Leeuwenhoek first observed microorganisms from the plaque on his teeth. He is commonly referred to as the Father of Microbiology for his discoveries.

ca 1665: The Englishman Robert Hooke created a compound microscope and coined the word "cell" for organisms.

1793: Edward Jenner, an English physician, carried out the first attempt at vaccination. He discovered that an infection with cowpox, the mild disease, protected one against the deadly smallpox.

1857: Louis Pasteur (Figure 3), a French chemist and microbiologist, demonstrated that fermentations were microbial processes. Thus began the Golden Age of Microbiology, which lasted for the next 60 years. Pasteur is commonly regarded as the Father of modern microbiology.

Figure 3: Louis Pasteur

1861: Louis Pasteur proved that bacteria were able to cause diseases. He published a theory called the germ theory of disease.

1864: Mr. Pasteur used brilliant experiments to prove that only existing microbes could give rise to other microbes. This confirmed theories of Ibn Al-Khatib, Fracastoro, and others. It also ended the spontaneous generation theory, which people believed for millennia. Louis Pasteur also created the word microbiology.

1869: Friedrich Miescher, a Swiss biologist and physician, discovered nucleic acids (Deoxyribonucleic acid or DNA, specifically), which he called nuclein at the time.

ca. 1878: Julius Petri, Robert Koch's assistant, developed glass dishes for culturing (growing) bacteria.

1881: Robert Koch, a German scientist, grew bacteria in individual forms called colonies. This technique of getting colonies of a single bacterial type is called pure culture.

1882: Robert Koch used agar to produce a pure culture of Mycobacterium tuberculosis, which causes tuberculosis. Fanny Hesse, the wife of Koch's assistant Walther Hesse, suggested the use of agar. He expanded Pasteur's work on the germ theory of disease. He developed some rules called Koch's Postulates to determine the microbe responsible for an infectious disease.

1884: Ernst Abbe, Carl Zeiss, and Otto Schott formed a company in Germany which improved the microscope and created the modern optical microscope.

1884: Hans Christian Gram developed a staining process using dyes, called Gram staining.

1885-1901: Albrecht Kossel isolated and described the four bases that form DNA: adenine, guanine, cytosine, and thymine, in addition to uracil, found only in RNA.

1887: Sergei Winogradsky, a Ukrainian microbiologist, discovered that some bacteria could use chemicals to produce energy.

1892: Dmitry Iwanowski showed that something that could pass through a special filter (unlike bacteria) caused a disease in tobacco plants called tobacco mosaic (Figure 4).

Figure 4. Tobacco Mosaic Virus

1898: Martineaus Beijerinck called Iwanowski's filterable agent a virus, studied viruses, and thus created the area of virology.

1880-1900: Several scientists, including students of Pasteur and Koch, discovered many bacteria.

1910: A coworker of Koch's, Paul Erlich, and his team created a chemical called Salvarsan. It could kill the bacteria that caused syphilis without killing the human cells surrounding it. He called it a "magic bullet", and it was the first targeted chemotherapy.

1923: David Bergey, an American bacteriologist, established the Bergey's Manual, a very important reference manual in microbiology. Scientists still use the manual in the identification of bacteria, to date.

1928: A Scottish scientist, Alexander Fleming, discovered that the mould

Penicillium, growing in pure bacterial cultures, killed surrounding bacteria by producing a substance he named penicillin.

1933: Ernst Ruska, a German, developed the transmission electron microscope (TEM). They make clearer images of microorganisms by using a beam of electrons instead of light, and this results in images with a higher resolution.

1937: Manfred von Ardenne developed the scanning electron microscope (SEM), which produces high-resolution, 3D images (Figure 5). It works by scanning a sample's surface with a focused beam of electrons and providing information about its surface shape and composition. These microscopes are useful in viewing smaller microorganisms on which the optical microscope cannot focus.

1938: Warren Weaver coined the term **molecular biology**. The field is a mix of research in biology, chemistry, and physics. Molecular biology is aimed at providing explanations to life.

1940: American scientists Howard Florey and Ernst Chain mass-produced Penicillin.

1940-1948: Selman Waksman, an American microbiologist, discovered streptomycin and other chemicals from soil bacteria. He called them antibiotics, and he became known as the Father of antibiotics.

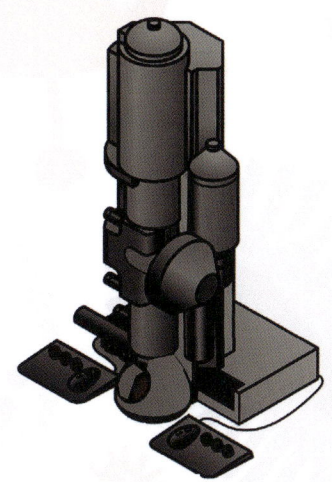

Figure 5. An electron Microscope

1943: The second Golden Age of Microbiology began when Italian Microbiologist Salvador Luria and German physicist Max Delbruck discovered that *Escherichia coli*, a gut microbe, could spontaneously mutate to resist infections by viruses. This observation caused many biologists to jump on the bandwagon and try to use microbes to answer important research questions in all of biology.

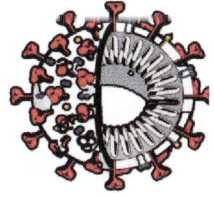

KNOWING MICROBES

1943: Mary Hunt, an American laboratory technician, obtained a mouldy cantaloupe from the market, and isolated the *Penicillium chrysogenum* strain that produced large amounts of penicillin. She became known as "Mouldy Mary" (Figure 6).

Figure 6: Mary Hunt

1944: Oswald Avery and others showed that DNA was the genetic material in bacterial cells. They also showed that DNA was the basis for heredity.

1952: Alfred Hershey and Martha Chase confirmed the claim by Avery and others about DNA using a virus that infects bacterial cells.

1953: James Watson and Francis Crick reported the double-helix structure of DNA (Figure 7).

Cytosine
Guanine
Adenine
Thymine

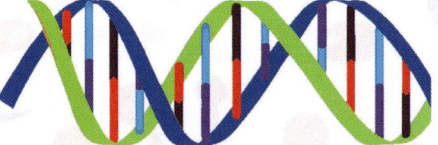

Figure 7. The Chemical Structure of DNA

1957: Arthur Kornberg discovered the first DNA polymerase, the enzyme that copies DNA, from *E. coli*.

1958: Francis Crick showed how DNA makes proteins in *E. coli*.

1962: Roger Stanier and Cornelius van Niel divided all organisms into procaryotes that do not have a cell nucleus and eucaryotes that do.

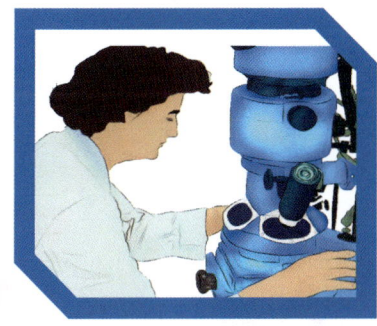

1964: June Almeida (Figure 8), a Scottish pioneer in virus imaging, was the first to identify the first coronavirus. Her work aided the understanding of the SARS-CoV-2 virus responsible for the COVID-19 pandemic.

Figure 8: June Almeida

1969: Thomas Brock isolated a bacterium called *Thermus aquaticus* from Yellowstone National Park in the United States of America (USA). Its polymerase, called *Taq* polymerase, could withstand high temperatures.

1973: Scientists Herbert Boyer and Stanley Cohen created the first genetically modified organism (GMO). They inserted foreign genes into *Escherichia coli*. Other transgenic animals and crops have since followed.

1977: Carl Woese, an American microbiologist and biophysicist, and coworker George Fox, discovered a group of unicellular procaryotes. They called these organisms Archaea and placed them in a third kingdom different from plants and animals.

1977: Frederick Sanger developed the first DNA sequencing method that made it possible to read the exact order in which the nucleotides are arranged within DNA.

1980: The United States Supreme Court ruled that living human-made microorganisms can be patented. This ruling resulted in the commercialisation of biotechnology.

1984: Luc Montagnier and Robert Gallo discovered that the Human Immunodeficiency Virus (HIV) caused a disease called Acquired Immune Deficiency Syndrome (AIDS).

Figure 9: Kary Mullis

1985: Kary Mullis (Figure 9), an American biochemist, invented the Polymerase Chain Reaction (PCR). PCR is a process that can make several millions of copies of DNA in a short time using the *Taq* polymerase. This invention appears to be the beginning of the Third Golden age of microbiology. PCR and molecular biology

1987: Yoshizumi Ishino and coworkers discovered unusual, repeated sequences while studying the sequences of *E. coli*.

have made it possible to understand microbes, their communities, what they do, and how they evolve.

1990: Carl Woese drew the Tree of Life commonly accepted today based on molecular characteristics and placed Archaea as a separate Domain of Life.

1998: Stewart Cole and coworkers decoded the DNA sequence of Mycobacterium tuberculosis, a bacterium that claims more lives than any other infectious agent.

2000: Francisco Mojica and Ruud Jansen discovered more repeated sequences in Archaea and named them CRISPR. Jansen also discovered CRISPR-associated systems of enzymes called Cas (Cas-9 is one of the most popular).

2005: Roche developed Next-generation sequencing technologies. This technology was several times faster than existing methods.

2012: Jennifer Doudna and Emmanuelle Charpentier discovered that the microbial CRISPR-Cas9 could be used to deliberately edit the genome. This procedure is one of the most important biological discoveries of all time. CRISPR is a technology that will likely change life, and thus the world, as we know it. They won the 2020 Nobel Prize in Chemistry for their discovery of these genetic scissors.

2016: Swedish scientist Stefan Jansson served the first known meal made from CRISPR-modified crops: tagliatelle with CRISPRy-fried cabbage.

2020: Barney Graham, Kizzmekia Corbett, Kathrin Jansen, Ugur Sahin, and Ozlem Tureci led teams that independently developed vaccines against the deadly SARS-CoV-2 virus, which caused the COVID-19 pandemic.

Today: **You** are next! How will you contribute to the advancement of microbiology?

Observing Microorganisms: Microscopy

OBSERVING MICROORGANISMS: MICROSCOPY

To view most microorganisms, we need to grow them first. Since microbes are found in all materials, all equipment and substances used in growing microbes must first be sterilised. Sterile petri dishes are half-filled with specific culture media. The samples containing the microbes we want to view are commonly diluted in a sterile liquid such as distilled water and transferred to the media in the dishes. These dishes are then stored (or incubated) at ideal temperatures. After some time, discrete colonies begin to form on the media (Figure 10).

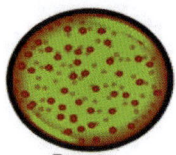 

Bacteria　　　　　Fungi

Figure 10. Colonies of bacteria and fungi growing on media in petri dishes

Microorganisms are so tiny that our eyes need some assistance to view them. Firstly, scientists place a part of a colony onto a glass slide and stain them in various ways, such as the Gram staining technique, and view them with a microscope (Box 1).

Microscopy, the use of microscopes to study microorganisms, is the basis of microbiology. In the 1660s, Anton van Leeuwenhoek, a Dutch draper, designed simple instruments which are the first known microscopes.

─────── KNOWING MICROBES ───────

Box 1: Viewing bacteria with an optical microscope

- Wear the appropriate laboratory protective equipment, especially a laboratory coat and safety goggles

- Smear the bacterial colony in the center of a glass slide, stain and cover with a cover slip. Place a drop of microscope oil on to the slip.

- Place the slide between the clips on the microscope stage

- Move the slide to the centre and directly under the smallest objective

- Move the stage up and down to focus the sample using the coarse and fine adjustment knobs

- Adjust the head until you can see through the eyepieces to the slide conveniently without squinting

- Once you can see the sample, select the most appropriate objective for your use

- Take notes and draw images of what you view, or use a camera to take a micrograph, where available

He used them to observe microorganisms in water and his mouth. He called the microbes *"dierkens"* or little animals. In 1674, Mr. van Leeuwenhoek became the first person to observe bacteria with these microscopes, which were illuminated by sunlight.

Optical or light microscopes create large images of small objects, i.e., magnify them, using light and glass lenses (Figure 11). These images are, however, not very clear at high magnifications because light has a long wavelength. The shorter the wavelength, the clearer the image.

Figure 11. A microbiologist using an optical microscope

Nowadays, advanced microscopes do not use light but use electrons. Since electrons' wavelength is a thousand times shorter than visible light, they can capture smaller details and provide more excellent resolution (Figure 12). The world's most powerful electron microscope is currently located in Cornell University in the USA. It works by shooting a beam of a billion electrons per second, at a material and can probe as deeply as the space between atoms!

 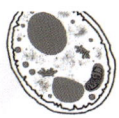

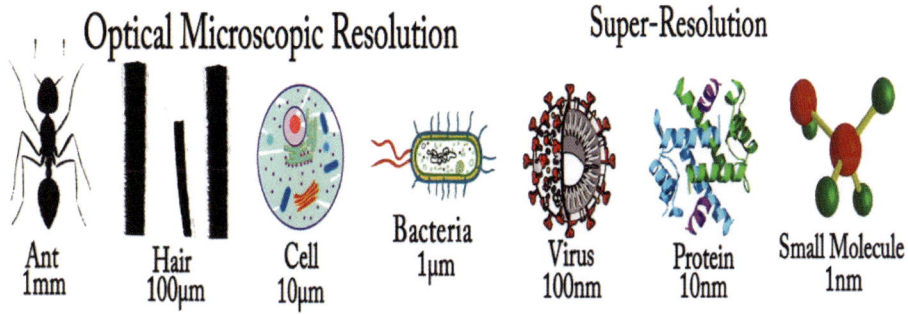

Figure 12: Microscopic resolution levels

Currently, an estimated 99 % of bacteria and archaea cannot be grown in the usual way by culturing in the laboratory! To study them scientists use new advanced techniques based on DNA sequencing including metagenomics.

More powerful optical and electron microscopes are also being developed and this is helping our understanding of microbes. In the near future, we would probably have a microscope app on our phones.

Classification of Microorganisms

CLASSIFICATION OF MICROORGANISMS

Over 3 billion years ago, unicellular microorganisms became the first forms of life to exist on Earth. However, microorganisms can change rapidly, so they have adapted and evolved to survive in different environments. Today, they are ubiquitous, present in every environment, including the Arctic and inside volcanoes! Scientists estimate that there are over 5×10^{30} bacteria alone in the world. That is 5,000,000,000,000,000,000,000,000,000,000 bacterial cells!

Microorganisms are diverse and can exist in unicellular (one-celled) forms, multicellular (many-celled) forms, or as cell clusters. The arrangement of the genetic material of the microorganisms is a basis for classification. The genetic material is loosely arranged inside the cell in microbes called prokaryotes. On the other hand, it is enclosed in the nucleus in eukaryotes.

This classification is now replaced with the newer and well-accepted three-Domain system. This system begins with a microbe called the "Universal Ancestor" and runs in levels from the Domain, the largest and most-inclusive group, to the species, the smallest and most specific group. Each classification level is further split into smaller, more related groups.

KNOWING MICROBES

THE THREE DOMAIN SYSTEM

The top and largest classification group is called the Domain, which determines if an organism is a procaryote or a eucaryote. Microbiologists now classify all microorganisms into three Domains, namely Archaea, Bacteria, and Eukarya (Figure 13). All members of Archaea and Bacteria are unicellular procaryotes. Similarly, all members of the Eukarya Domain are (you guessed it) eucaryotes. Some members of Eukarya are multicellular, including yourself. Each Domain is divided into Kingdoms, the second-largest group.

The Kingdom in the Domain Archaea is known as Archaebacteria, while that of the Domain Bacteria is Eubacteria. Similarly, the microbial Eukarya Kingdoms are:

- Protista/Protoctista: made up of protozoa and algae
- Fungi: made up of yeasts, moulds, and mushrooms.

Each Kingdom is further split into six lower-order groups (known as taxa), ending with species.

.

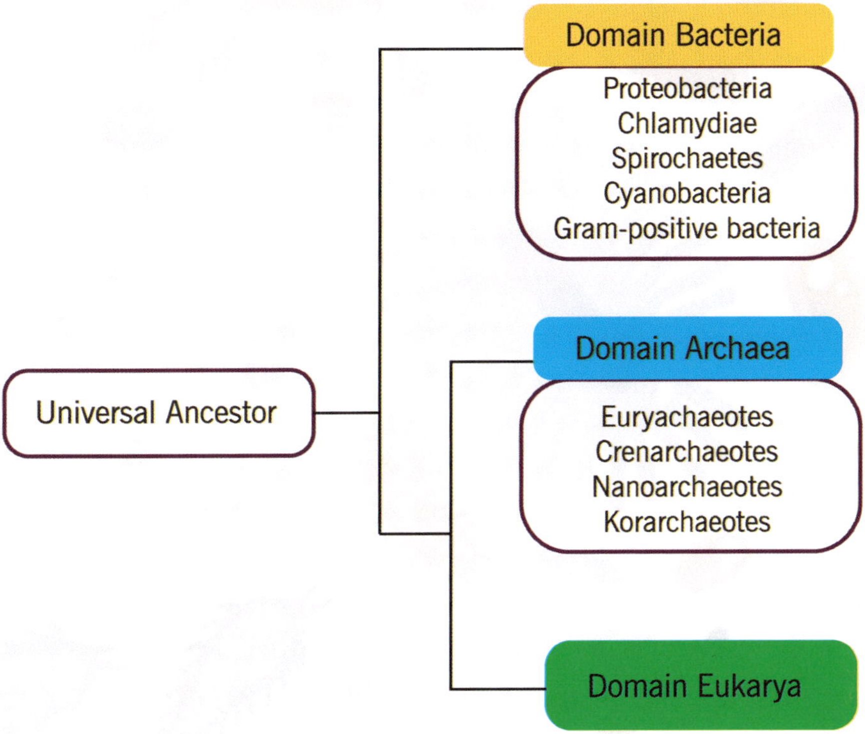

KNOWING MICROBES

```
                        ┌─── Domain Bacteria
                        │      Proteobacteria
                        │      Chlamydiae
                        │      Spirochaetes
                        │      Cyanobacteria
                        │      Gram-positive bacteria
                        │
                        │─── Domain Archaea
Universal Ancestor ─────┤      Euryachaeotes
                        │      Crenarchaeotes
                        │      Nanoarchaeotes
                        │      Korarchaeotes
                        │
                        └─── Domain Eukarya
```

Figure 13. A tree showing how the three Domains are related and the major microbial phyla they contain

All microorganisms have two names in a system known as binomial nomenclature. These names are the genus and species names. The names are in Latin and are written in italics or underlined.

Here is an example of the complete taxonomic classification of a bacterium:
Domain: Bacteria
Kingdom: Bacteria
Phylum: Proteobacteria
Class: Gammaproteobacteria
Order: Enteriobacteriales
Family: Enterobacteriaceae
Genus: *Escherichia*
Species: *Escherichia coli.*
The first letter of the genus name is capitalised it and can be shortened to just the first letter, e.g., *E. coli.*

THE SIX MICROBIAL GROUPS

The Domain system is not commonly used outside academic circles. Microbes from the three Domains are frequently separated into common groups. Let us look at the six common types of microbes in detail.

1. Algae

The Algae are one half of the Kingdom Protoctista of the Domain Eukarya. They are found in damp areas, oceans, and rocks. They range in size from microscopic forms such as *Micromonas* to giant forms like kelps, which can reach a skyscraper's height! They contain chlorophyll or other pigments and are thus able to photosynthesise (Figure 14).

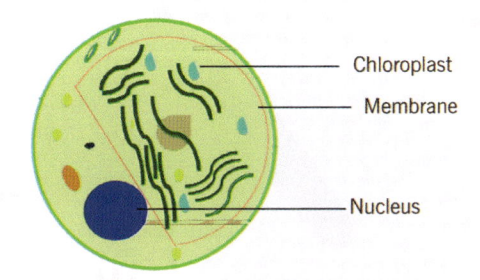

Figure 14. A photosynthetic alga, *Chlorella vulgaris*

2. Archaea

Archaea are a Domain of unicellular procaryotes that separated from "true bacteria" early in the evolution of life. So, while they are similar to bacteria, they do not have peptidoglycan in their cell walls. They are found in extreme environments where there is no oxygen. Suppose an environment is acidic, salty, boiling, or freezing. In that case, archaea will be the most likely microorganisms found having a field day there! They are also found in the guts of humans and animals. Archaea have not been linked to diseases in humans.

3. Bacteria

The Domain Bacteria contains the true bacteria are the most numerous microbes around us, with several million species belonging to this Domain (Figure 15). Bacteria are procaryotes and have a thick wall outside their cell called the cell wall.

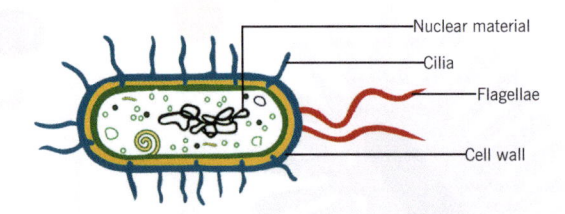

Figure 15. A typical bacterial cell

Bacteria are unicellular and exist in one of five main shapes: bacillus (rod-like), coccus (spherical), spirillum (spiral), vibrio (curved) and spirochaete (corkscrew). (Figure 16).

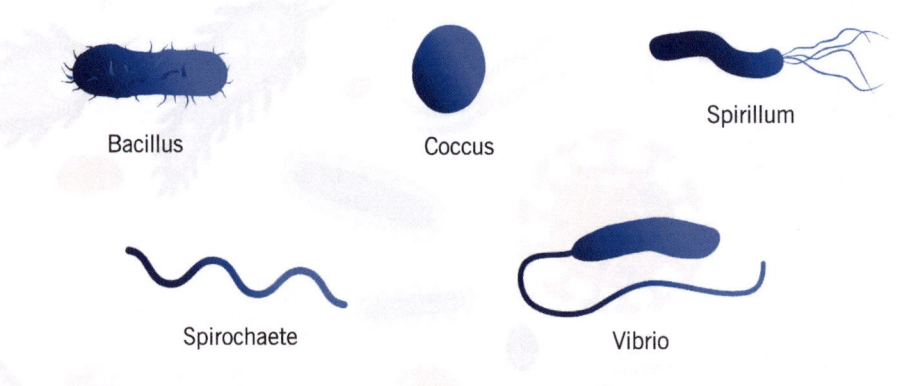

Figure 16. Different bacterial shapes

Most bacteria have a complex material called peptidoglycan in their walls, while some do not. The amount of this material in the cell wall partly determines what colour bacteria stain when treated with dyes. Bacteria with thick peptidoglycan layers stain purple, and microbiologists refer to them as Gram-positive bacteria. Those with thin layers, and another exterior membrane stain pink and are called Gram-negative.

Some common bacteria are *Staphylococcus aureus* and *Escherichia coli*, both of which are found in our bodies, and only occasionally cause diseases. Another common bacterium is *Lactobacillus delbrueckii*, which microbiologists and food scientists use to make yoghurt. *Clostridium botulinum* produces the toxin botulinum toxin used in cosmetic applications as botox.

4. Fungi

Fungi are eucaryotes; there may be up to 1.5 million species in nature, existing mainly as moulds, yeasts and mushrooms. Moulds include the fungi you see in some old, damp houses or the black powder you may find on spoiling onions which is *Aspergillus niger* (Figure 17).

Penicillium chrysogenum

Saccharomyces cerevisiae

Aspergillus niger

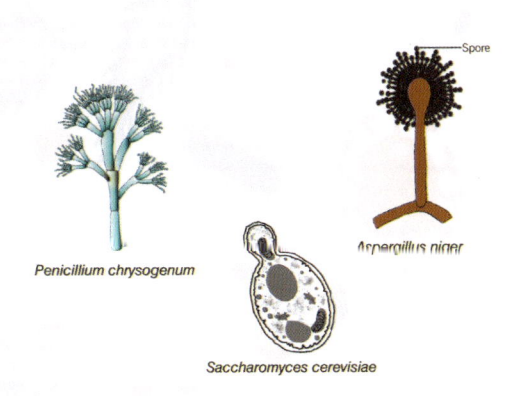

Figure 17. Common fungal species

Some moulds like *Aspergillus fumigatus* can make you sick, while others like *Penicillium chrysogenum* are used to produce antibiotics. Yeasts are unicellular fungi and common examples include *Saccharomyces cerevisiae* used to make bread and wines, and *Candida albicans*, which causes disease. Mushrooms are the fruiting bodies of some kinds of fungi. Which is your favourite mushroom?

5. Protozoa

Protozoa (singular, protozoon) are the other half of the Kingdom Protoctista and are unicellular eucaryotes that exist as parasites in plants and animals. Common examples are *Entamoeba histolytica* and *Plasmodium falciparum*, which cause amoebiasis and malaria, respectively (Figure 18).

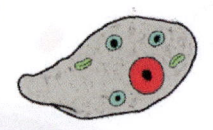

Figure 18. The protozoon *Entamoeba histolytica*

6. Viruses

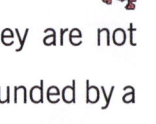

Discussing viruses can be slightly tricky. This is because they are not exactly cells but are particles made of a nucleic acid core surrounded by a protein coat. Thus, while many scientists classify viruses as microorganisms, they are not actually considered to be alive. Confusing, right? This status is because, in some ways, viruses are like living things, e.g., in their ability to reproduce. However, unlike other living things, they cannot reproduce on their own, only inside a living cell. Viruses can infect living cells and use the cells' machinery to reproduce, destroying the cells and causing diseases and death. They also cannot carry out metabolism on their own.

The most (in-)famous virus of this century, so far, is SARS-CoV-2 (Figure 19). This virus causes COVID-19, the devastating pandemic disease which originated in Wuhan, China, in 2019.

Figure 19. The Severe Acute Respiratory Coronavirus-2 (SARS-CoV-2)

How Microorganisms Move

HOW MICROORGANISMS MOVE

Do you feel the need to follow the scent of food to the kitchen or run to hug your loved ones when you see them after some time apart? In the same way, microorganisms also move towards favourable stimuli such as nutrients. Again, like we need to run away from harm, microorganisms also need to avoid harmful chemicals. Bacteria also seek each other out, cluster, and move together like dancers in a show (Box 2) using their various structures (Figure 20).

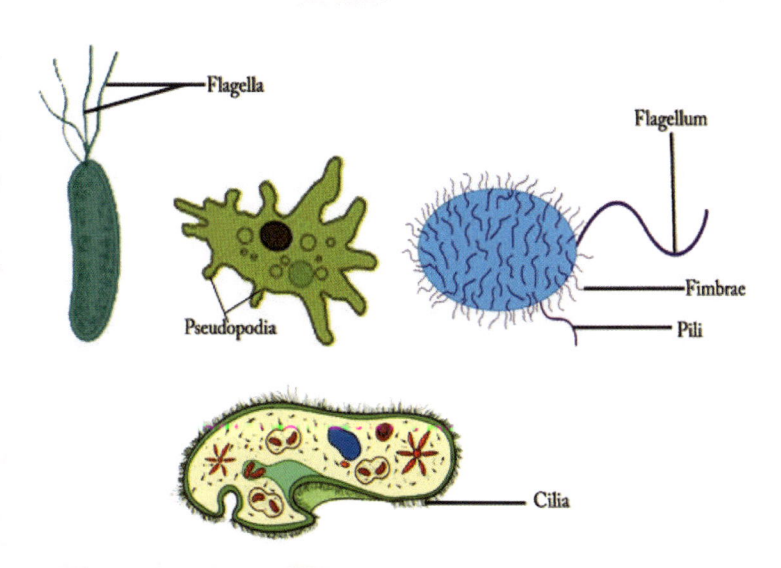

Figure 20. Different locomotion structures in microbes

KNOWING MICROBES

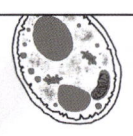

Box 2: How microorganisms move

Bacteria have many different structures that enable movement.

- Some bacteria like *Salmonella enterica* can slide under certain conditions.

- Others use flagellae which look like whips and work like a boat's propeller to push the bacterium forward.

- Pili are used to stick to a surface and pull the bacteria forward when pulled inwards; . They are also useful in reproduction.

- Cilia are hair-like extensions that beat together in a coordinated manner that propels protozoa such as *Paramecium* sp. through water.

- Fungi do not move, but are fixed in the materials on which they grow.

Microorganisms Are A Part Of Us!

MICROORGANISMS ARE A PART OF US!

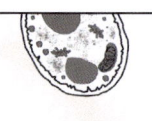

As you have come to realise now, microorganisms are a part of all aspects of our lives. Humans acquire microorganisms from birth and live with them throughout life. Without vaccination, many children would die in infancy from diseases. Similarly, pathogenic microbes will be able to kill millions annually if we did not have antibiotics. Sadly, today's improper use of antibiotics means that many of these pathogens are no longer killed by antibiotics, meaning that they have developed antibiotic resistance.

While we mainly know microbes as causes of illnesses, most of the time, they live "peacefully" in all parts of our bodies. Different parts of our bodies host entirely different communities of microorganisms, and we actually depend on them and their activities to keep us safe. Microorganisms also exist in water, and some like *Vibrio cholerae* and *Entamoeba histolytica* can make us very ill.

There are over 1,800 microbe types in the air we inhale. Researchers estimate that we breathe in a million microbes daily! Do not hold your breath now; most of the airborne microorganisms are harmless. However, some like *Alternaria* species can cause allergies, while some diseases like tuberculosis and influenza are spread through the air.

Microbes are most numerous in the soil, especially in warm, moist soils.

Microbes, especially fungi, are essential in the decomposition of dead matter. Imagine if fallen tree logs and dead animals remained in the environment forever? Gross!

MICROORGANISMS AND DISEASE

With all the microbes living in, on, and around us, there are likely to be consequences. As mentioned earlier, some bacteria can make us sick. These germs are scientifically known as pathogens, and they cause illnesses known as infectious diseases. Humans acquire microorganisms in several ways.

Contact

Contact is the most common way that we acquire pathogens. Pathogens can be directly passed on to us when we have contact with carriers. Infectious skin diseases are spread this way. Similarly, the SARS-CoV-2, influenza, HIV, and hepatitis viruses could be transmitted to us by shaking hands, kissing, and having sexual intercourse with sufferers of these diseases. We can also indirectly pick up some pathogens when we touch our shoes, doorknobs, toilet seats, etc. We even inhale many microorganisms from the environment, especially spores of fungi like *Aspergillus fumigatus* and *Blastomyces dermatitidis*.

Contaminated food and water

When we eat raw or improperly cooked foods and unwashed fruits, the microbes in the foods enter our digestive system. Transmission is also possible when we drink unpurified water such as seawater or water contaminated with sewage. This is why eating foods like sushi, rare steaks, unwashed berries and vegetables, and raw eggs can result in diseases. The most common of these foodborne diseases are diarrheal diseases such as cholera caused by *Vibrio cholerae*, typhoid fever caused by *Salmonella* sp., and *E. coli* diarrhoea. We should avoid uncooked foods or obtain them only from licensed food sellers.

Bites

Some creatures known as vectors can transmit disease-causing microbes to humans. Insects are common vectors, and they pass microbes to us through bite wounds. Mosquitoes are the most common vectors of diseases in humans. They can transmit the parasite *Plasmodium* which causes malaria, through their bites. Similarly, they also transmit the viruses that cause yellow fever, Zika and Dengue diseases. Ticks can transmit the bacterium *Borrelia burgdorferi*, which causes Lyme disease. Lastly, fleas can transmit *Yersinia pestis,* the bacterium that causes the plague!

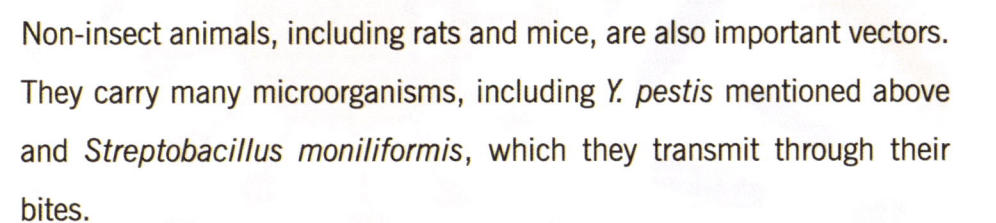

Non-insect animals, including rats and mice, are also important vectors. They carry many microorganisms, including *Y. pestis* mentioned above and *Streptobacillus moniliformis*, which they transmit through their bites.

COMMON INFECTIOUS DISEASES AND THEIR SYMPTOMS

One is said to have an infection when bacteria, viruses, fungi, or other pathogens find their way inside the body and multiply in number. The symptoms of an infection depend on the microbe type and the affected body part. Table 1 shows some common diseases caused by microorganisms.

Table 1. Some common infectious diseases and their characteristics.

Disease	Route of transmission	Infectious agent	Affected area	Symptoms
Impetigo	Contact	*Staphylococcus aureus* or *Streptococcus pyogenes*	Skin	Reddish, itchy sores; blisters; crusts; fever
Boils	Contact	*Staphylococcus aureus*	Skin	Red and tender bump; pus; swollen lymph nodes; fever
COVID-19	Contact; Inhalation (airborne)	SARS-CoV-2	Respiratory system, and most organs: heart, lungs, brain	Loss of smell and taste senses; fever; breathing problems; organ failure; death
Ebola fever	Contact	Ebola virus	Entire system	Fever; chills; diarrhoea; muscular pain; bloody vomit; organ failure; death
Zika fever	Aedes mosquito bites (and contact)	Zika virus	Eyes, muscles, skin	Fever; rash; joint pain; red eyes; birth defects in babies of infected mothers
Aspergillosis	Contact (inhalation)	*Aspergillus fumigatus*; *A. terreus*	Lungs, bloodstream, brain	Breathing problems; fever; bleeding in the lungs; death
Salmonella infection/ Typhoid fever	Contaminated food, e.g., eggs	*Salmonella enterica*	Abdomen	Diarrhoea (bloody); vomiting; fever; abdominal pain;

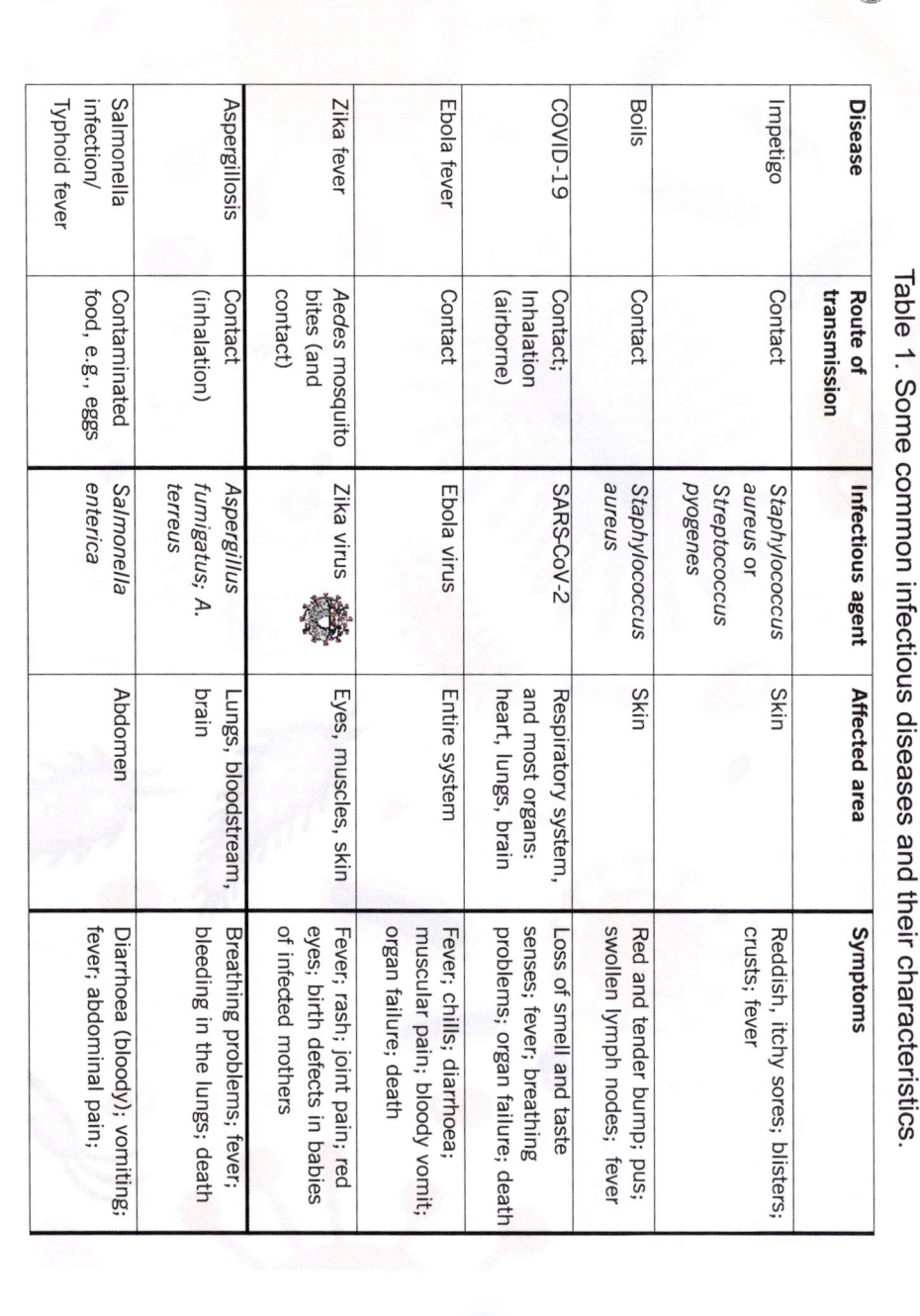

HOW MICROORGANISMS BENEFIT US

At this point, you may be thinking that microbes are just terrible nuisances. Well, you are only partially correct. Microorganisms are a part of many beneficial processes in our lives and contribute to our daily existence in more ways than you can imagine. The following are some ways in which microorganisms are a beneficial part of us.

Food

Microbes have been a part of our culinary experience from the beginning of history. Natural food preparation processes involving microorganisms are called fermentation. Humans have used yeasts, moulds, and bacteria for thousands of years to make foods and today, there are over 3,500 types of fermented foods that keep for longer and taste even better!

Food fermentation is widely practised in Africa, with most ethnic groups having their unique fermented foods. For instance Cassava is a tropical tuber that is rich in starch but also contains toxic chemicals. Africans ferment cassava to remove these toxins and preserve the cassava, they then use it to produce *fufu, lafun,* and *garri*. The main organisms involved are lactic acid bacteria (LAB) and some yeasts.

Additionally, Africans ferment corn and other grains using natural LAB and other microbes to produce *ogi*, a liquid food similar to custard, and *injera*, a tangy bread. Many condiments are made through fermentation, such as *ogiri, iru, ugba*, which are prepared from oilseeds and beans by bacteria like *Bacillus* species. In Africa, fermented drinks like *amahewu* and *burukutu* made from grains, *amasi* made from milk, and palm wine made from palm tree sap. *Unqombothi* is a South African beer made from fermented grains. Yvonne Chaka Chaka immortalised it in the timeless Afro-pop song of the same name she performed in 1988. Meat and fish fermentation are less common in Africa.

In Europe, beer, wine, and bread are among the most common fermented food products consumed. They are all made using mainly *Saccharomyces cerevisiae*, one of the most important microorganisms on Earth! Yoghurt and cheese are made by the fermentation of milk by LAB and are also popular fermented foods (Figure 21). These bacteria produce acids that give yoghurt its sour taste. They also produce long chains of sugar called polysaccharides, which gives yoghurt its thick texture.

A malt drink   Bread   Cheese

Figure 21. Some popular fermented foods

Meat fermentation is popular in Europe, with salami, pepperoni, chorizo, and dried ham produced by fermentation. Europeans popularly prepare sauerkraut by the fermentation of cabbage by naturally-occurring LAB.

In Asia, soy and fish are widely fermented. Common examples of fermented foods are tempeh, fungus-fermented soybeans; kimchi made from cabbage and other vegetables; *kombucha* made from black tea and sweeteners like honey; soy sauce; and miso made from soybeans and rice. A popular fermented food from South America is *atole agrio*, which is black maize fermented by LAB. Others include *curtido* and *kimchi*, which are similar to sauerkraut. In the Mediterranean/Middle East, *kushuk*, a dry food, *torshi* made from a vegetable mix, *doogh* a salted yoghurt, and *shanklish* a cheese covered with spices, are popular fermented foods.

Outside the fermentation of plant and animal products, some microbes themselves can be consumed as food as **single-cell protein** or microbial protein. Quorn® is a protein-rich meat substitute product made from *Fusarium venenatum,* a soil fungus. The fungus is grown in large vats known as fermenters, dried, mixed with egg albumin and pressed into the shape of meat products.

Are you hungry yet? That would be completely understandable!

Health

The **microbiome** is the total body of microbes that live within and on our bodies. Immediately a baby is born, it begins to collect microbes on all parts of its body and this process continues throughout life. This mighty but invisible luggage of microorganisms we carry on and in us influences our health. We know microbes cause diseases, but it is important to note that they also support our health. That's right, microbes protect us from falling ill. Different groups of microbes live in the different parts of the human body and perform various activities. We carry the largest section of our microbiome in our guts (Figure 22).

There is a balance in the numbers and types of microbes in the human body, depending on one's age, lifestyle or genes. The "good" bacteria control the number of harmful microbes that can produce harmful poisons. This is why live yoghurt is a very healthy food, as the bacteria in it, *Lactobacillus*, *Streptococcus* and *Bifidobacterium* species, help maintain the right gut microbiome balance. This balanced microbiome is vital for our health, as if it is seriously changed, it can cause severe illnesses such as cancers or even death. This imbalance can be caused by the abuse of antibiotics, which could kill harmless microorganisms and leave dangerous ones like *Clostridium difficile* that causes severe illness.

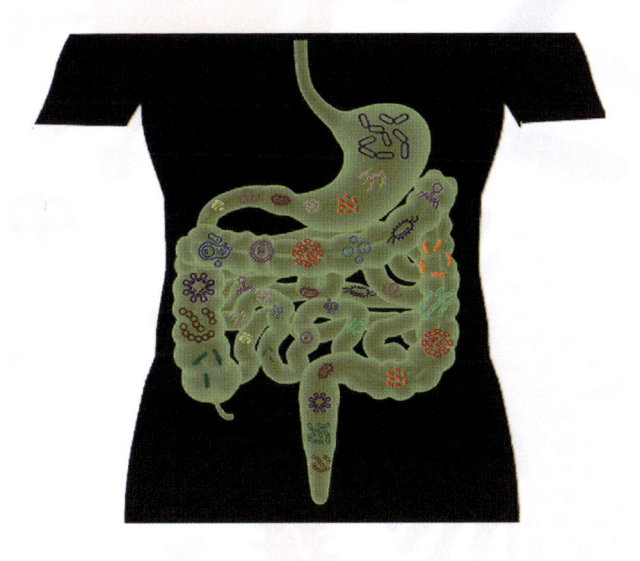

Figure 22. Microorganisms in the human gut

Additionally, microorganisms are used to develop vaccines, which protect us from diseases like pertussis, influenza, measles and even COVID-19. The organisms are killed or rendered harmless and then introduced into our bodies as vaccines. The body then puts up an immune response (effectively fighting a shadow) and develops the ability to protect us when we are confronted with live pathogens. Vaccines have helped to wipe out poliomyelitis which causes crippling and smallpox, a lethal disease.

Income

You read that right. Microorganisms are a massive source of income for many people in ways other than their roles in food production and vaccine production. For instance, citric acid is an organic acid found naturally in lemons and limes and used to flavour pop/soda drinks and many foods. Today it is almost entirely produced by growing *Aspergillus niger* on carbohydrate solutions. Using microorganisms to produce materials useful in industries is called industrial microbiology. Currently, industrial microbiologists produce antibiotics like streptomycin using *Streptomyces griseus*, chemicals like itaconic acid using *Aspergillus terreus*, and fuels like ethanol using *Zymomonas mobilis*. These and several other industries generate trillions of dollars annually.

Life

Apart from the microbiome, other organisms help to ensure that life literally goes on. The photosynthetic algae and bacteria even produce oxygen, which keeps life on Earth going. Fungi, and bacteria like *Staphylococcus, Bacillus, Pseudomonas,* and *Clostridium* species, are essential in the decay of dead plants and animals.

Additionally, bacteria belonging to the genera *Thiobacillus*, *Nitrosomonas*, and *Nitrobacter* recycle elements like sulphur, iron, carbon, and nitrogen. They help ensure that these vital elements are removed from decaying matter and made available for other life forms.

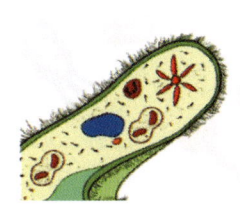

KNOWING MICROBES

Microorganisms in Extreme Environments

MICROORGANISMS IN EXTREME ENVIRONMENTS

What makes an environment too extreme for life? Excessive levels of factors such as acidity, salt, and radiation. You would consider a temperature of 50 °C extremely hot, right? Sun exposure at that temperature would cause your skin to burn severely! Extremophiles are organisms that have a "love for" and can survive and thrive in environments other organisms find too harsh. They are found in all three Domains of life, but most belong to the Archaea.

Many extremophiles have more than one "superpower". For instance, many alkaliphiles are also halophiles and are thus called haloalkaliphiles. They are found in the Mono Lake in the USA and the East African Rift Valley. *Deinococcus radiodurans* is the heavyweight champion of extremophiles as it can survive low temperatures, low pressure, acid, and dehydration! This is in addition to being the most radiation-resistant organism known. It is recognised by The Guinness Book of World Records® as the toughest bacterium. Box 3 gives some fun facts about extremophile microbes.

Box 3: Common classes of extremophiles and representative species

Acidophiles: These microbes love acidic environments. Yikes! An example is Helicobacter pylori which lives in the stomach with hydrochloric acid present. It causes stomach ulcers in humans.

Alkaliphiles: These love alkaline environments. An axample is Bacillus pseudofirmus found in the hyperalkaline spring in Zambales, Philippines.

Barophiles: These are only found in high-pressure environments such as the deep sea. A Moritella species was found in the Mariana Trench, nearly 11 km deep into the sea!

Halophiles: These organisms tolerate high amounts of salt. Examples are Salicola marasensis which was isolated from solar salterns in Peru, and Debaryomyces hansenii which is found in cheeses.

Thermophiles: These organisms can survive in very hot conditions, even above the temperature of boiling water! Remember Thermus aquaticus. It is a commonly-known thermophile.

Psychrophiles: These organisms, such as Pseudomonas syringae, love cold temperatures and can live in ice.

Radiophiles: These organisms have powers that put them in the same league as the Martian Manhunter and Superman. They can survive the extreme irradiation found in outer space and nuclear facilities.

CONCLUSION

CONCLUSION

You can see that humans have been curious about microorganisms for millennia. Through the work of various scientists, our understanding of them has increased. We now know that these incredible and fascinating organisms are incredibly numerous and they are not solely disease-causing agents as they are widely known. They are involved in everything from food to chemical production and have been classified based on their different characteristics and their roles for better understanding. We now know a lot about microorganisms, from how they move, how they cause diseases, how we can benefit from them, to their ability to live in extreme environments. Now microbiologists can readily identify several species using simple and more advanced techniques. However, there is still a lot to learn from and about them. Most microorganisms cannot be grown in the usual way, and scientists are working on complex molecular biology and other methods to identify and grow them.

Knowing microbes is an exciting and enlightening process, and we have only just scratched the surface. If you would you like to know more about these fascinating creatures, chat a microbiologist up.

GLOSSARY

GLOSSARY

3D: the property of having three dimensions, e.g., a box.

Agar: or agar-agar, is a jelly-like substance obtained from red algae. Amber: this is a resin from ancient trees that thickened and then became fossilised.

Archaea: a Domain of unicellular microbes that are similar to bacteria and are commonly found in extreme environments.

Atom: the smallest unit into which matter can be divided and still have the properties of the individual element, and without the release of electrically charged particles.

Bacteria: unicellular microbes that make up the Domain Bacteria, a large group of procaryotes.

Chemotherapy: the use of any drug to treat any disease. Nowadays, the word chemotherapy ("chemo") refers to the use of drugs to manage cancer.

Chlorophyll: a green pigment that plants, cyanobacteria, and algae use to produce their food from sunlight. Class: a taxonomic rank below Phylum and above Order.

CRISPR: Clustered Regularly Interspaced Short Palindromic Repeats. Portions of bacteria's sequence that provides them with immunity against viruses. Culinary: of cooking, or related to cooking.

Colony: a distinct group of bacteria or fungi growing on a solid agar medium.

Deoxyribonucleic acid/DNA: a complex organic molecule that carries the genetic information of all life. It is made of two strands that twist on each other to form a double helix.

Disease: an abnormal condition that negatively affects an organism and which is not caused by external injury.

DNA sequencing: the process of determining the order (sequence) in which the component building blocks (nucleotides) are arranged in a strand of DNA.

Domain: the highest taxonomic rank of organisms in the current Woese three-domain system of taxonomy.

Electron: a negatively charged subatomic particle that is diffracted like light.

Evolution: evolution is the gradual change in characteristics of a species over time. It is based on the idea that all species are related and gradually but continuously change over time.

Extremophile: an organism that thrives in an extreme environment.

Family: a taxonomic rank below Order and above Genus.

Gene: the basic unit of heredity. Genes are a sequence of nucleotides; segments of DNA.

Genera: plural of Genus.

Germ theory of disease: the theory that infectious diseases are caused by the invasion of the body by microorganisms (infection).

Genome: an organism's complete set of genetic instructions needed for that organism to exist, grow, and develop.

Genus: a taxonomic rank below Family and above Species. It is the first part of the two-part name given to organisms and is usually italicised.

Gram staining: a microbiological technique used to differentiate bacteria into two phrnad groups. This procedure distinguishes bacteria into Gram-positive and Gram-negative groups, which are stained violet/blue or pink/red, respectively.

HIV: Human Immunodeficiency Virus; the virus that causes AIDS.

Italics: A style of printing in which the letters lean to the right. The biological names of organisms are written in italics (or italicised) in print, which is equivalent to underlining them in writing.

Kingdom: the second highest taxonomic rank, just below Domain.

Lactic acid bacteria/LAB: a large group of bacteria used in food fermentations and include *Lactobacillus, Lactococcus, Streptococcus, Bifidobacterium,* and *Pediococcus*, among others.

Medium: (or growth medium) a solid or liquid material from chemicals and foods prepared to support the growth of microorganisms in the laboratory

Metabolism: all the chemical reactions involved in maintaining the living state in an organism.

Metagenomics: the study of genetic material collected directly from environmental samples.

Microbiome: the total body of microbes that live within and outside our bodies.

Micrograph: a photograph taken with the aid of a microscope.

Microbe: a living thing that exists all around us but which we are only able to see with the aid of a microscope

Microbiology: the area of science which involves the study of microorganisms.

Microscope: an instrument that makes a larger image of a small object too small to be seen with the unaided eye.

Millennia: the plural of "millennium" - a period of 1000 years.

Nucleus: a membrane-bound organelle that contains genetic material (DNA) of eukaryotic organisms.

Nucleic acid: a complex organic substance present in cells, such as DNA. The molecules consist of many nucleotides linked in a long chain. Nucleotides: the basic building block or unit of nucleic acids such as DNA.

Order: a taxonomic rank below Class and above Family.

Organism: any living thing that has an organised structure, can react to stimuli, reproduce, grow and adapt.

Pathogen: a microorganism that causes disease.

Petri dish: a flat, shallow transparent dish and lid made of glass or plastic used to culture microorganisms.

Phylum/Phyla: a taxonomic rank below Kingdom and above Class.

Probiotic microorganisms: these are live bacteria and yeasts that, when consumed, can provide benefits such as improved immunity.

Protists: group of unicellular procaryotes which are neither plants, animals, nor fungi. They include protozoa, unicellular algae, and slime molds

Protozoa: unicellular, free-living, or parasitic members of the Domain Eukarya.

Resolution: the shortest distance between two points on a specimen that can still be recognised by the eye or camera system as separate entities. A term used to describe the ability of a microscope to distinguish detail.

Sequence: a DNA sequence is the order in which the four base components in DNA are arranged.

Single-cell protein: also known as microbial protein, this refers to the dried form of edible microorganisms, including algae, bacteria, fungi, and cyanobacteria.

Species: the basic and lowest unit of classification. A species is a grouping of organisms with similar characteristics. They can (potentially) breed with each other to produce fertile offspring. It is the second part of the two-part name given to organisms and is usually italicised.

Spore: thick-walled cells produced by many bacteria and fungi for reproduction and as protection against harsh environments.

Sterilisation: the process of using heat or radiation to destroy all microorganisms on the surface of an article or in a fluid.

Taxonomy: the science of naming and classifying organisms into groups/ranks based on biological characteristics.

Vaccination: the act of introducing a vaccine into the body to protect one from a specific disease.

Vaccine: a substance that helps protect against certain diseases.

Vector: a creature that transmits disease-causing microbes to humans.

Virology: a part of microbiology concerned with the study of viruses and virus-like agents.

Wavelength: The distance between two identical adjacent points in a wave.

NOTES

Printed in Great Britain
by Amazon